MIND OF ITS OWN: CLYDE COMES ALIVE!

By

Sue Glover

ISBN: 0-7596-7891-X

This book is printed on acid free paper.

Printed in the United States of America

1stBooks - rev. 04/18/02

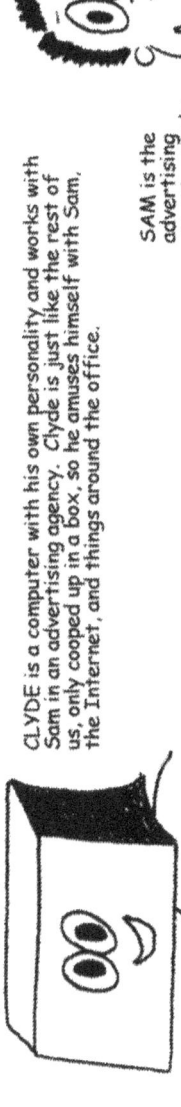

CLYDE is a computer with his own personality and works with Sam in an advertising agency. Clyde is just like the rest of us, only cooped up in a box, so he amuses himself with Sam, the Internet, and things around the office.

SAM is the advertising expert, or so he thinks. He's actually the "lovable loser" kind of guy that has trouble finding dates and dressing appropriately so he relies on Clyde's advice, which you can bet will leave him in a pickle.

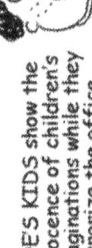

JOE'S KIDS show the innocence of children's imaginations while they terrorize the office.

MARGE, Sam's secretary, is much brighter than Sam and usually helps him get out of a jam (or sometimes helps him get into one as well).

JOE is Sam's boss, but he's pretty confused and his kids play havoc in the office.

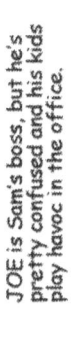

1

BIRD (Guest appearance)

GUS is the office custodian and usually cleans up Sam's messes. He tends to overreact and take his job too seriously, but in a humorous way, of course.

HANS (MR. FIX-ALL, INC.) is the computer repairman who tries really hard to fix things, but doesn't quite have the knack. Because of his antics, Hans is on a first-name basis with the 911 operator.

MIKE is the computer mouse and is always on the lookout for cheese.

DR. WACKO is the company psychologist and makes cameo appearances to check up on the team's sanity.

3

4

5

6

7

10

11

14

19

21

22

25

29

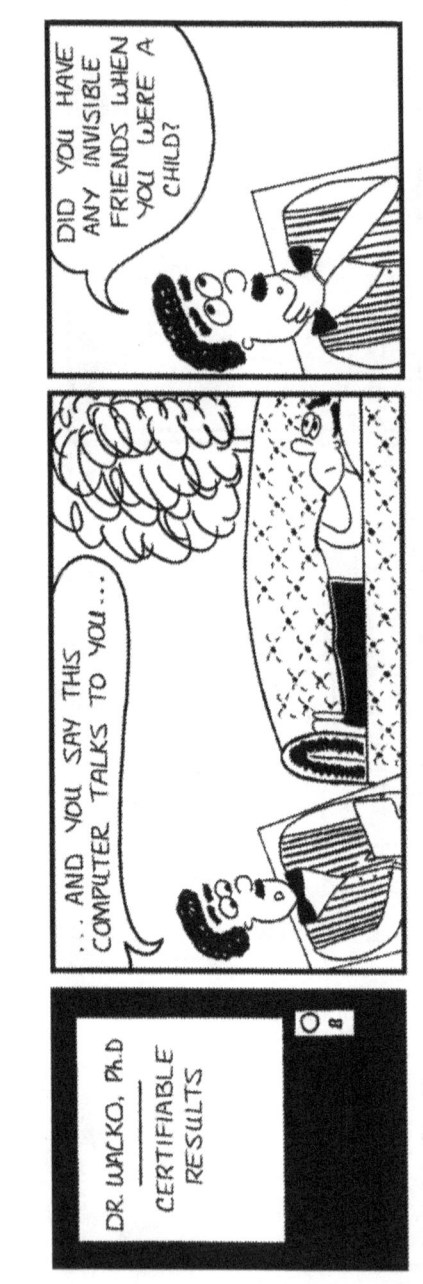

31

33

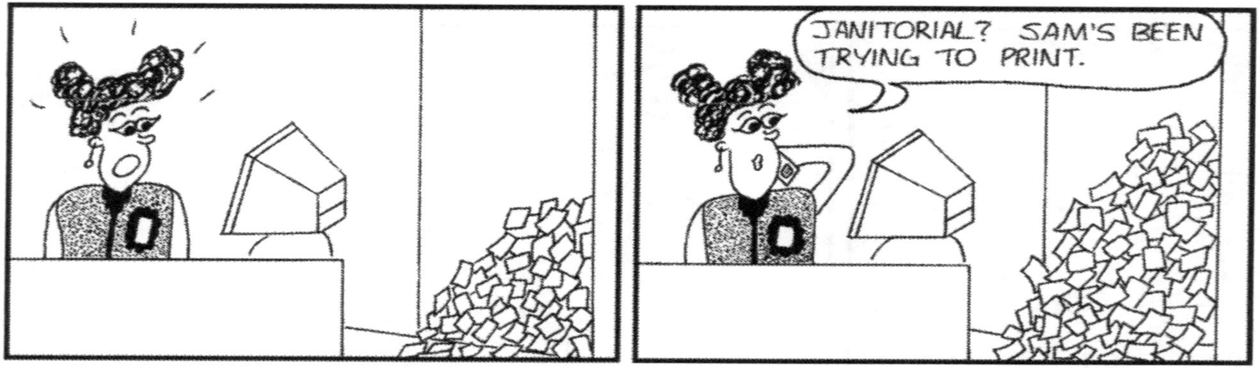
JANITORIAL? SAM'S BEEN TRYING TO PRINT.

35

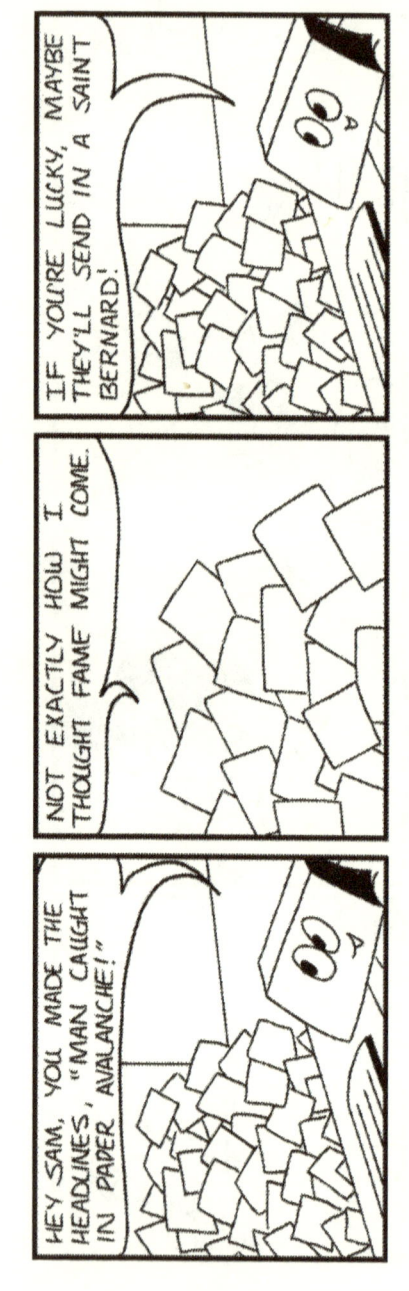

39

43

WHAT KIND OF GAMES DOES IT HAVE?

WOULDN'T YOU RATHER GO PLAY WITH THE DRINKING FOUNTAIN?

HOW ABOUT TIC-TAC-TOE?

THERE'S A GREAT FICUS TREE DOWN THE HALL.

OH, THIS WILL GO OVER GREAT WITH THE BOSS!

HEY - IT WORKED, DIDN'T IT?

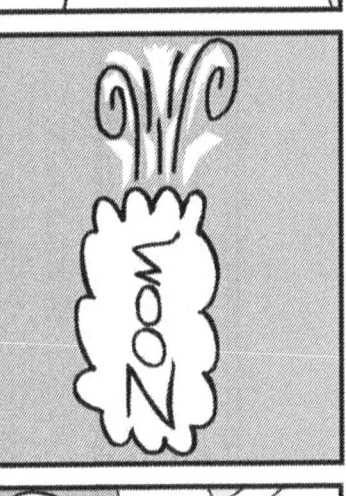

ZOOM

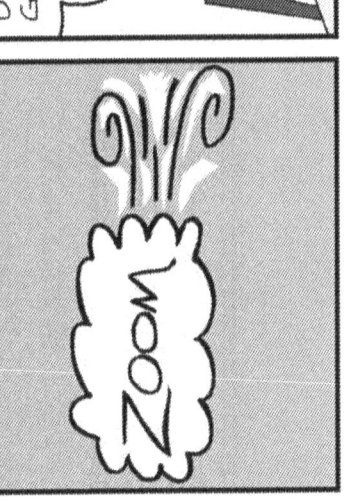

BOO!

AYYYHHH!!

49

51

55

59

63

OKAY—YOU'RE HOOKED UP TO THE NET.

GREAT, HOW DOES IT WORK?

FIRST YOU PRESS THIS SWITCH, THEN TYPE TOOL...

MR. ALL FIX INC.

HANS

...THEN HIT THAT KEY, THIS KEY, ENTER YOUR BIRTH DATE, ENTER YOUR MOTHER'S BIRTH DATE, HIT THAT KEY...

HANS

WHAT WAS THAT?

I GUESS I GOT SOME WIRES CROSSED.

I HATE IT WHEN THAT HAPPENS.

MR. ALL FIX INC.

HANS

POOF

...WAIT A COUPLE OF MINUTES WHILE IT...

87

89

91

95

97

101

103

107

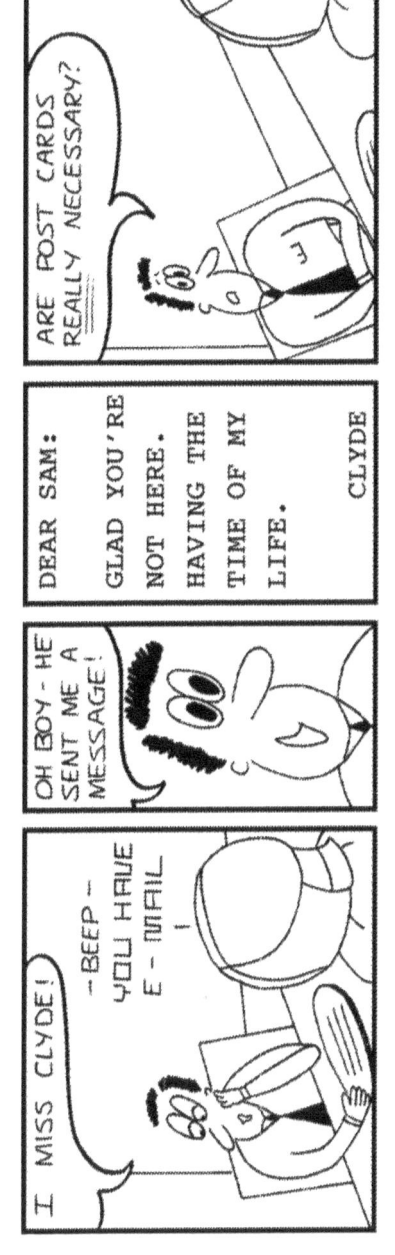

OKAY CLYDE — TWO CAN PLAY THE POSTCARD GAME!

REPLY:

DEAR CLYDE:

DOING **GREAT** WITHOUT YOU. DON'T HURRY BACK!

SAM

THAT OUGHT'A MAKE HIS CIRCUITS SOGGY!

REPLY TO REPLY:

DEAR SAM:

OKAY — SEND MONEY.

CLYDE

119

125

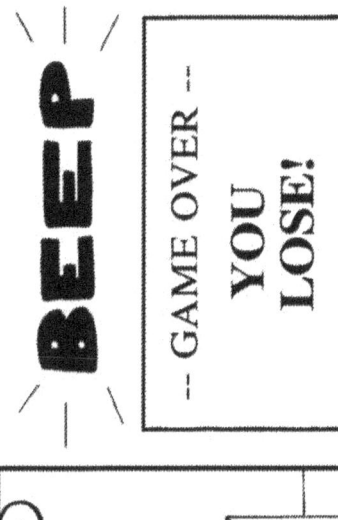

129